Rafaella Gouveia Mendes
Leila Leal da S. Bonetti
Vania A. Nascimento

Germination of Annonas seeds

Rafaella Gouveia Mendes
Leila Leal da S. Bonetti
Vania A. Nascimento

Germination of Annonas seeds

Breaking dormancy with gibberellic acid

ScienciaScripts

Imprint
Any brand names and product names mentioned in this book are subject to trademark, brand or patent protection and are trademarks or registered trademarks of their respective holders. The use of brand names, product names, common names, trade names, product descriptions etc. even without a particular marking in this work is in no way to be construed to mean that such names may be regarded as unrestricted in respect of trademark and brand protection legislation and could thus be used by anyone.

Cover image: www.ingimage.com

This book is a translation from the original published under ISBN 978-613-9-67688-0.

Publisher:
Sciencia Scripts
is a trademark of
Dodo Books Indian Ocean Ltd. and OmniScriptum S.R.L publishing group

120 High Road, East Finchley, London, N2 9ED, United Kingdom
Str. Armeneasca 28/1, office 1, Chisinau MD-2012, Republic of Moldova, Europe
Printed at: see last page
ISBN: 978-620-7-70481-1

Every choice, no matter how small, is a form of seed that we weed in the flowerbed that we are. One day, everything that we now silently plant, or allow to be planted in us, will be a plant that can be seen from afar...

Father Fabio de Melo

INDEX

PREFACE

This book is the result of academic work carried out by the authors at the University of Minas Gerais, Ituiutaba Unit. It is divided into three chapters which report on the effects of gibberellic acid (GA_3) on three species of the Annonaceae family. The fruit trees analyzed were *Annona cacans (*araticum-cagao), *A. muricata* (graviola) and *A. squamosa* (pinha), which have economic, food, medicinal and ecological importance.

Annonaceous plants have seeds that are difficult to germinate due to dormancy, a phenomenon of physical or physiological origin presented by some species, which has abscisic acid as the phytohormone that inhibits or delays the process. Therefore, gibberellin acts in opposition to abscisic acid in seed dormancy. Dormancy is considered both a problem and an alternative to prevent seeds from germinating, even in favorable humidity, temperature, light and oxygen conditions. It can be considered a problem because it greatly reduces the natural germination of species and it can be a strategy because it allows species to increase their survival rates in the environment. Thus, germination only occurs when the external and internal conditions of the seeds are the most suitable for the establishment of seedlings. However, as this phenomenon can last for days, months or even years, using techniques that help overcome dormancy is important as an alternative to increasing germination rates and achieving rapid and homogeneous germination.

One technique used to break seed dormancy is the use of GA3, one of several types of gibberellins, which are phytohormones widely distributed throughout the plant. Seed germination may require gibberellins to activate embryo growth, break down and mobilize endosperm reserves in the presence of water and cancel dormancy in some species.

GA3 is an important inducer of the germination process in seeds, especially those that need to go through stratification, i.e. a period of cold to germinate.

The seeds of the species studied were subjected to standardized tests described in BRASIL (2009) in order to assess their quality, germination capacity, the influence of GA3 and various substrates on germination. The aim of all these tests was to find ways of obtaining a greater quantity of cherimoya seedlings, an alternative crop for small producers in the tijucana region.

The araticum-cagao has fruits that are important for feeding birds, reptiles and mammals, its natural dispersers, and are less appreciated by humans because they cause diarrhea (CARVALHO, 2003). Considered a pioneer plant, it grows quickly in full sun and is recommended for reforestation.

Graviola is the second most cultivated anonaceous plant in Brazil, especially in the states of Alagoas, Ceara, Paraiba, Pernambuco, Para and Minas Gerais, with strong growth in the south of Bahia, as an alternative to cocoa farming. There is little *fresh* consumption of graviola because the fruit is very perishable. Most of the harvest is sold in the form of graviola paste, which consists of the fruit peeled and frozen. Pulps, juices, nectars and other products are sold in agro-industries (LEMOS, 2014).

The fruit of the pine tree is usually eaten *fresh.* The main producing state in Brazil is Bahia, but there are plantations in Alagoas, Pernambuco, Sao Paulo and Ceara. The semi-arid areas of the northeast and the north of Minas Gerais are where pine cone cultivation has increased the most. Drier and warmer regions are more suitable for the development of the crop, as they provide favorable conditions for phytosanitary management (LEMOS, 2014).

Breaking dormancy in annonaceous seeds is the subject of many studies, since their germination is low and their fruit is economically important. Expanding research and encouraging the cultivation of fruit trees, including cherimoya, based on knowledge of physiology and the application of chemical substances can increase productivity and at the same time provide dignity and income generation for producers, strengthening the local agro-industry.

CHAPTER I

GERMINATION AND VIGOR OF ARATICUM SEEDS INFLUENCED BY GA3 IN DIFFERENT SUBSTRATES

GERMINATION AND VIGOR OF ARATICUM SEEDS INFLUENCED BY GA3 IN DIFFERENT SUBSTRATES

ABSTRACT: Native species of the Cerrado are little used in the recovery of degraded areas due to the lack of knowledge about their physiology, ecology, propagation techniques and management. Among them is the araticum-cagao *(Annona cacans* WARM.), from the Annonaceae family,

with propagation and cultivation hampered by the low germination rate of the seeds. The aim of this study was to evaluate the germination and vigor of araticum-cagao seeds under the effect of gibberellic acid (GA_3), in different substrates. To assess seed germination and vigor, four treatments were carried out with gibberellic acid on seeds obtained commercially and collected in the municipality of Ituiutaba-MG. After soaking, the physiological quality of the seeds was assessed using the Water Content, Electrical Conductivity, Germination and Germination Speed Index tests in different substrates. Analysis of the results shows that the concentration of 1000mg of GA_3 had a positive influence on seed germination and vigor and that the substrates paper and sand with vegetable soil were suitable for germination tests.

KEYWORDS: Araticum-cagao; phytoregulator; seed physiology.

INTRODUCTION

The Cerrado ecosystem is considered to be one of the richest savannas in the world. In its vastness, there is variability in climate, soils and diversity of fauna and flora (SILVA *et al.,* 1994, RIBEIRO; WALTER, 1998).

The high rate of deforestation serves as a warning of the importance of creating reforestation measures in order to reverse or mitigate this process. It is necessary to understand the physiology of native plant species and promote seed germination, monitor plant development, identify ideal patterns for reproduction, all of which will certainly contribute to reforestation support with plants originating from the ecosystem. In general, native representatives in the recovery of degraded areas are little used due to the lack of knowledge about the physiology, ecology, propagation and management techniques of these species (CARPI *et al.,* 1996; RANIERI *et al.,* 2003).

The *Annonaceae* family comprises around 120 genera and more than 2,000 species, with the araticum-cagao *(Annona cacans* WARM.) being one of its main representatives (CARVALHO, 2003). It is a deciduous, pioneering plant, and because it grows quickly in full sun, it is recommended for reforestation support.

The araticum-cagao has fruits that are not very popular with humans due to their diarrhea-like taste, but these fruits are part of the diet of birds, reptiles and mammals, which are its natural dispersers (CARVALHO, 2003).

The impermeability and dormancy of seeds naturally protect them and provide them with conditions for survival during periods that are unsuitable for their establishment and development. While the sensitivity of seeds allows them to awaken automatically when the environmental conditions are right.

The propagation and cultivation of araticum is made difficult by the low germination rate of the seeds, which lasts up to a year (RIBEIRO *et al.,* 2000; SALOMAO *et al.,* 2003). CARVALHO (2003) says that araticum-cagao seeds take between 30 and 113 days to germinate and have an average germination rate of 30%.

Dormancy, common in native species, is a phenomenon whereby seeds do not germinate even under favorable environmental conditions due to endogenous factors determined by the seed itself (MARCOS FILHO, 2005, CARVALHO; NAKAGAWA, 2000), requiring pre-germination treatments.

Rizzini (1976) and Carvalho (2003) suggest that araticum-cagao is dormant due to embryonic indifferentiation. In this type of dormancy, the seeds, after dispersal, require a longer time to germinate because they contain immature embryos, with still incomplete development.

Reports on breaking the dormancy of seeds of various species can be found in scientific literature, indicating the use of mechanical, chemical and thermal scarification methods.

Araticum's dormancy is not broken by methods such as heat shock in hot water, prior soaking of the seeds in natural water, or mechanical scarification of the seed coat. However, the latter contributes to the absorption of gibberellic acid and, consequently, to seed germination and seedling emergence (PEREIRA *et al.,* 2004). They also report that treatment with gibberellic acid breaks the dormancy of araticum seeds, with immersion for four days at concentrations of 250 to 2000 mg/L or for two days at concentrations of 1000 to 2000 mg/L being more effective.

Gibberellic acid (GA_3), one of the more than 100 compounds known as gibberellins, stimulates a-amylase and other hydrolytic enzymes, promoting the hydrolysis of reserves stored in the seed. GA_3 is a phyto-regulator which, among its many effects, promotes the breaking of seed dormancy in many seeds and stimulates germination. However, its dosage and handling are important factors to consider, not only because it is a product with a high commercial value, but also because it can have an inhibitory effect depending on the concentration (Castro *et al.,* 2005).

Studies with plants of the *Annona* genus have demonstrated the efficient use of gibberellins in overcoming dormancy, a characteristic of this genus. METIVIER (1986) highlighted the role of gibberellins in germination, acting both to break dormancy and to control the hydrolysis of reserves. They are indicated in pre-germination treatments.

According to Abreu *et al.* (2005), even if the embryos are fully mature, attention must be paid to the type of substrate, as they are essential for supplying the nutrients needed for plant growth and development. It also provides ideal conditions for the germination process.

An ideal substrate should have structure, consistency, sufficient porosity to drain excess water and maintain adequate aeration near the root system, a good capacity to retain water and not contain toxic substances, inoculations of diseases, invasive plants and other pests. According to Fonseca (2002), the choice of substrate should be based on its availability, economic viability, standardization and homogenization, with little variation in physical and chemical characteristics.

It is therefore important to choose a substrate that is compatible with the requirements of the seed, such as size, shape and tolerance to desiccation. BRASIL (2009) recommends using sand as the substrate for the germination test because it produces good results for seedling production. In the case of cerrado species, soil from the cerrado itself can be used, as well as other options, including the use of vegetable soil, which results from the decomposition of parts of the plant, such as leaves, stems, fruit and others.

The physiological quality of seeds must be assessed in terms of viability and vigor. Viability determines whether the seed is alive or dead. Vigor represents attributes of physiological quality that are not revealed in the germination test. It is determined under stress conditions or by measuring the decline of some biochemical or physiological function (NAKAGAWA, 1999); in short, it expresses the level of energy that a seed has available to carry out the tasks of the germination process (CARVALHO; NAKAGAWA, 2000).

Many tests can be used to assess seed germination and vigor, including germination and germination speed index, water content and electrical conductivity. According to Brasil (2009), seed germination in a laboratory test is the emergence and development of the essential structures of the embryo, which demonstrate its ability to produce a normal plant under favorable field conditions.

Water content has a direct influence on various aspects of the physiological quality of seeds, which is why it is determined in official seed lot quality tests. It is related to the ideal harvest period, seed weight and susceptibility to damage from heat, freezing, fumigation, mechanical damage or pests (GRABE, 1989). High water content can affect seed quality not only during the storage period, but also during processing operations, making handling more difficult and reducing the efficiency of the machinery used in the processing processes (CARVALHO; NAKAGAWA, 2000).

The electrical conductivity test makes it possible to estimate seed vigor. Its principle is based on the aging process of the seed, with deterioration and consequent loss of integrity of the cell membrane systems, which become more permeable, allowing electrolytes to leach out. The change in electrical resistance caused by the leaching of electrolytes from the seed's tissues into the water in which it was immersed (VIEIRA and KRZYZANOWSKY, 1999) can be measured using a conductivimeter. Thus, the extent of cell membrane disorganization can be estimated by the amount of solutes leached from seeds soaked in distilled water.

From the germination tests, it is possible to evaluate the effectiveness of the substrate and the concomitant influence of gibberellic acid (GA_3), in different concentrations, to improve the germination process of Araticum-cagao and verify the maximum expression of its vigor.

The aim of this study was to evaluate the germination and vigor of araticum-cagao seeds under the influence of GA_3, in different substrates and under laboratory and field environmental conditions.

METHODOLOGY

Some of the *Annona cacans* WARM. seeds were purchased commercially through the IBflorestas website, while others were collected from mother trees in the rural area of Ituiutaba-MG, in July 2016. The seeds were extracted manually from the fruit, washed and placed on paper towels to dry in the open air, in the shade, for 24 hours. Before the experiments, all the seeds were aseptically soaked in 2% sodium hypochlorite for 10 minutes. They were then washed in distilled water and soaked for 48 hours in their respective treatments containing different concentrations of GA3. The seeds were washed again in distilled water and placed in seedbeds containing the substrates and on germitest paper, moistened daily throughout the experiment.

In order to evaluate the germination of Araticum- cagao seeds, eight treatments were carried out with the following concentrations of gibberellic acid (GA_3): treatment 1, control 0 $mg.L^{-1}$; treatment 2, 500 $mg.L^{-1}$; treatment 3, 1000 $mg.L^{-1}$; treatment 4, 2000 $mg.L^{-1}$. For each treatment, the following concentrations were used.

Four different substrates were used in the treatment: sand and vegetable soil, sand, cerrado soil and germitest paper.

The seed germination experiments on the substrates were conducted in accordance with Brasil (2009), outdoors, in natural environmental conditions in the city of Ituiutaba-MG, located in a region of the cerrado biome. The germination test on germ paper was conducted in a controlled germination chamber at 25°C. The number of germinated seeds was assessed daily, with the criterion for emergence being the exposure of part of the seedling on the surface of the substrate. Electrical conductivity, germination speed index (GVI) and germination percentage were evaluated.

The germination percentage was assessed for 50 days, while the GVI was assessed 10 days after germination. For each treatment, the GVI suggested by Biruel *et al.* (2007) was calculated, which shows the number of seeds germinated per day, where GVI = G1/N1+ G2/N2 +... Gn/Nn, where G1, G2... Gn equals the number of seeds germinated, and N1, N2... Nn corresponds to the number of days.

The electrical conductivity test was carried out according to HAMPTON; TEKRONY (1995) and VIEIRA; KRZYZANOWSKI (1999). The seeds were soaked in plastic cups (200 mL) containing 75mL of the GA3 solution from each treatment and kept in controlled conditions at 25°C.24 h^{-1} . After the soaking process, the electrical conductivity was read using a DIGIMED conductivity meter, model CD 21, with a constant 1.0 electrode and the results were expressed in "micro Siemens per centimeter per gram" (pS cm^{-1} g).$^{-1}$

The results were analyzed using a completely randomized design, with five replicates per treatment. Statistically, the measures of dispersion and analysis of variance (ANOVA) were calculated, applied to the factorial scheme and the means were compared using Tukey's test at 5% probability.

RESULTS AND DISCUSSION

Table 1 shows the average percentage moisture content of *Annona cacans* WARM. seeds subjected to the standard oven water content test (105°C/24h).

Table 1. Average values of the water content test (TA) of *Annona cacans* WARM seeds.

	WATER CONTENT (%)	
TREATMENT	**COMMERCIAL**	**COLLECTED**
T1	30,784(±1,89) A	6,140(±0,56) A
T2	27,667(±1,84) A	6,600(±0,75) A
T3	23,388(±1,44) B	6,623(±0,42) A
T4	23,556(±1,26) B	6,405(±0,34) A

C.V (%)	10,08	9,97

Means of five repetitions. Values followed by the same letter in the column do not differ by the Tukey test ($p < 0.05$). Values in brackets correspond to the standard deviation. CV = coefficient of variation.

The seeds collected had a low percentage of moisture, between 6.140 and 6.623%, with no significant difference between the batches used for each treatment. The seeds obtained from the market showed a higher percentage of moisture in the lots from treatments 1, with 30.784%, and 2, with 27.667%, which differed significantly from the other lots. According to CARVALHO (2003), this is due to the recalcitrant behavior of the species, related to storage. GUSATTO *et al.* (2012), analyzing the breaking of dormancy in *A. cacans seeds,* obtained values that showed this behavior and observed a moisture content of 10.45% in the seeds, while DALANHOL (2013) found 16.5% moisture for the seeds of the same species.

Table 2 shows the results of the electrical conductivity test of *Annona cacans* WARM. seeds soaked for 24 hours in different treatments, given in pS cm^{-1} g^{-1} .

Table 2. Average values of the electrical conductivity (EC) test on *Annona cacans* WARM seeds. data in pS cm^{-1} g^{-1} .

	ELECTRICAL CONDUCTIVITY - EC (pS.cm'vg'1)	
TREATMENT	**COMMERCIAL**	**COLLECTED**
T1	9,105(±0,29) B	13,468(±0,99) B
T2	27,586 (±0,30) B	16,319(±0,79) B
T3	28,636 (±0,33) B	13,493(±1,26) B
T4	45,124 (±0,27) A	22,102(±0,88) A
C.V (%)	8,83	11,13

Averages of five replicates. Values followed by the same letter in the column do not differ by the Tukey test ($p < 0.05$). Values in brackets correspond to the standard deviation. CV = coefficient of variation.

In the electrical conductivity (EC) test, electrolyte leaching occurred in ascending order among the treatments, with T1 having the lowest value. Treatment T4 differed statistically from the others and showed less vigor (Table 2), suggesting that GA_3 at a dose of 2000mg negatively influenced the reorganization of cell membranes and led to a greater loss of exudates. TORRES *et al.* (2009) obtained similar results for sesame seeds, regardless of the combination, amount of water and temperature used, pointing to batch 4 as having low vigor. The results obtained in this study were similar between the seeds regardless of their origin.

The *Annona cacans* WARM. seeds were subjected to the IVG and germination percentage tests. The results shown in Table 3 refer only to the seeds collected, since those obtained from the market did not germinate.

Table 3. Average germination values (%) and germination velocity index (GVI) of *Annona cacans* WARM. seeds, submitted to treatments with GA3 and in different substrates.

TREATMENTS	SUBSTRATE	GERMINATION (%)	IVG
T1	Germitest paper	30,0 (±1,72) B	10,25 (±1,73) A
	Sand+Earth Vegetable	26.0 (±1.14) Ba	9,41 (±1,70) A
T2	Germitest paper	44,0(±1,31) B	12,02 (±1,58) A
	Sand+Earth Vegetable	38,0 (±1,48) B	10,24 (± 1,48) A
T3	Germitest paper	74,0 (±1,48) A	16,41(±1,61) B
	Sand+Earth Vegetable	72,0 (±1,28) A	12,73 (±1,49) A
T4	Germitest paper	52,0 (±1,67) B	12,77 (±1,69) A

Sand+Earth Vegetable	46.0 (±1.88) Ba	10,86 (±1,67) A
C.V (%)	7,34	8,11

Averages of five replicates. Values followed by the same letter in the column do not differ according to the Tukey test ($p < 0.05$). Values in brackets correspond to the standard deviation. CV = coefficient of variation.

Seed germination began at 92 days after sowing and ended at 136 days. Carvalho *et al.,* (2001) in an experiment with Araticum-do-brejo *(Annona glabra)* found that germination started at 55 days, while Almeida (1987), at 250 days after sowing. Estrela *et al.,* (2002) and Souza Neto *et al.,* (2005) considered the effectiveness of gibberellic acid in reducing the time required for germination to be promising.

The IVG showed better results in the seeds submitted to treatment 3, on germitest paper, which was the only one to differ significantly from the others. Souza Neto *et al.* (2005), studying *Annona glabra* L., stated that seeds soaked in gibberellic acid germinated more quickly.

The germination percentage showed that the seeds submitted to treatment 3 had the best results, reaching 74% on germitest paper and 72% in the sand substrate with vegetable soil, the only treatment that differed significantly. Gusatto (2012) observed 45% germination of *Annona cacans* WARM. seeds, while Braga Filho (2014) obtained 66% germination of *Annona crassiflora* M., values lower than those obtained in this experiment. The results obtained here indicate that the concentration of gibberellic acid, GA3, used in treatment 3, was suitable for breaking dormancy and stimulating germination of *Annona cacans* WARM.

Among the substrates analyzed for germination of araticum-cagao seeds, the best performance was seen in germitest paper, both in the germination percentage test and in the IVG. This can be explained by the controlled germination conditions, since according to Marcos Filho (2005) the promotion of germination basically depends on four factors: temperature, water availability, phytohormone application and the presence of light. The use of germ paper in a germination chamber controls all these factors.

Sand with vegetable soil showed a germination percentage of up to 72%, which makes it the best substrate for conventional cultivation of araticum-cagao. Pina-Rodrigues; Jesus (1990a) corroborate this observation when they state that high temperatures associated with substrates with high moisture retention can show good results in recalcitrant seeds, as is apparently the case with araticum. POPINIGIS (1985) states that the influence of the substrate on germination is due to its characteristics, such as structure, degree of aeration and water retention capacity, factors found in the combination of sand and topsoil.

The seeds collected showed good germination, but the seeds purchased commercially did not germinate. This may be due to the storage conditions, which were not informed by the supplier. Morais *et al.* (2009) observed the germination viability of
only six months in *Annona squamosa L.* seeds. Cruz *et al.* (2016), studying the same species, observed that the refrigerator environment provided a higher percentage of germination at 6 months of storage, while the natural environment provided twice the percentage of abnormal seedlings compared to the refrigerator environment, at 3 months of storage, reducing the percentage of germination.

In general, the use of gibberellic acid to break dormancy in *Annona cacans* WARM. seeds was efficient at a concentration of 1000 mg of GA_3 , providing a significant increase in the germination rate. This was also observed in seeds of *Annona crassiflora, Annona squamosa, Annona muricata and Annona cherimola* (BERNARDES *et al.,* 2007; CAVALCANTE et *al.,* 2007; DIAS, 2003; LIMA-BRITO *et al.*, 2006; STENZEL; MURATA; NEVES, 2003).

CONCLUSIONS

The results lead us to conclude that, for *Annona cacans* WARM., GA_3 at a concentration of 1000 mg had a positive influence on seed germination and vigor; that paper and sand with topsoil were the most suitable substrates for germination tests and that storage conditions are relevant when assessing the physiological quality of seeds.

REFERENCES

ABREU D.; NOGUEIRA A. C.; MEDEIROS A. C. S. Effect of substrate and temperature on the germination of cataia seeds *(Drimys brasiliensis* Miers. Winteraceae). **Revista Brasileira de Sementes**, vol. 27, n° 1, p.149157, 2005.

ALMEIDA, S. P.; SILVA, J. A.; RIBEIRO, J. F. **Food use of native Cerrado species: araticum, baru, cagaita and jatoba.** Planaltina: EMBRAPA-CPAC, 1987. 83 p. (EMBRAPA-CPAC. Documentos; 26).

BERNARDES, T. G. *et al.* Effect of storage and phytohormones on the physiological quality of araticum *(Annona cassiflora* Mart.) seeds. **Pesquisa Agropecuaria Tropical**, v. 37, n. 3, p. 163-168, sep. 2007.

BIRUEL R. P; AGUIAR I. B. de; PAULA R. C. de. Germination of pau-ferro seeds subjected to different storage conditions, chemical scarification, temperature and light. **Revista Brasileira de Sementes.** Associapao Brasileira de Tecnologia de Sementes, v.29, n.3, p.151-159, 2007. Available at <http://hdl.handle.net/11449/3627>. Accessed: 20.08.2018.

BRAZIL. Ministry of Agriculture, Livestock and Supply. Secretariat for Agricultural Defense. **Rules for Seed Analysis**. Brasilia: Mapa/ACS, 2009.

399 p. ISBN 978-85-99851-70-8

CARPI, S. M. F.; BARBEDO, C. J.; MARCOS-FILHO, J. Osmotic conditioning of *Cedrela fissilis* Vell seeds. **Revista Brasileira de Sementes**, Brasilia, v.18, n.2, p.271-275, 1996.

CARVALHO, J. E. U. de; NASCIMENTO, W. M. O. do; MULLER, C. H. Tolerance of araticum-do-brejo *(Annona glabra* L.*)* seeds to desiccation and freezing. **Revista Brasileira de Fruticultura**, Jaboticabal-SP, v.23, n.1, p.179-182, April 2001.

CARVALHO, N. M. de; NAKAGAWA, J. **Seeds: science, technology and production.** 5.ed. Jaboticabal: FUNEP, 590p. 2012. ISBN: 978-85-7805-090-0

CARVALHO, P. E. R. **Especies arboreas brasileiras.** Brasilia: Embrapa Informapao Tecnologica, v. 2, 2003. 8573831677, 9788573831672

CARVALHO, P. E. R. **Ariticum-Cagao**. Colombo: Embrapa Florestas, 2003. Circular Tecnica, 81, 8p., 2003.

CASTRO, P. R. C., KLUGE, R. A.; PERES, L. E. P. **Manual of Plant Physiology**: theory and practice. Piracicaba: Editora Agronomica Ceres, 2005, 650p.

CAVALCANTE, T. R. M.; NAVES, R. V.; BRAGA FILHO, J. R.; SILVA, L. B. Influence of substrates and seed storage on the emergence and growth of araticum plantlets *(Annonaceae).* **Bioscience Journal**, Uberlandia, v. 23, n. 4, p. 11-20, Oct./ Dec. 2007.

CRUZ, M. A *et al.*Storage of atemoya cv.'Geffner' seeds in natural and refrigerated environments. **Proceedings...**In: 1st International Meeting of Agricultural and Technological Sciences Crisis: technologies for overcoming challenges in the agricultural sector, Dracena: FEPAF, September 21-23, 2016, p.189- 194.

DALANHOL, S. J. *et al. Dormancy* in seeds of *Annona cacans* WARM. *(Annonaceae).* **Revista Academica: Ciencias Agrarias e Ambientais** (PUCPR. Impresso), v. 11, p. S183-S189, 2013.

DIAS, G. B. **Effect of different temperatures and gibberellic acid concentration on the germination of atemoya seeds** *(Annona cherimola* Mill x *Annona squamosa* L.) **Gefner cultivar**. 2003. 46 f. Course Conclusion Paper (Graduation in Agronomy) - State University of Western Paraná, Marechal Candido Rondon, 2003.

ESTRELA, C. T. *et al.* Sexual propagation of araticum (Annona crassiflora Mart.). In: CONGRESSO BRASILEIRO DE FRUTICULTURA, 17, 2002, Belem. **Proceedings...** Belem: SBF, 2002. 1 CD.

FONSECA, E.P. *et al.* Quality assessment of *Trema micrantha* (L.) Blume seedlings produced under different shading periods. **Revista Arvore**, Viposa, v. 26, p. 515-523, 2002.

GRABE, D. F. Measurement of seed moisture. In: STANWOOD, P. C. and McDONALD, M. B (Eds). Seed Moisture. Madison: **The Crop Science Society of America**, p.69-92, 1989.

GUSATTO, F. C. *et al.* Breaking dormancy in araticum cagao *(Annona cacans)* seeds. **Anais...** In: 4th Paranaense Forestry Congress, September

10-14, 2012.

HAMPTON, J.G.; TEKRONY, B.M.(Eds.). **Handbook of vigour test methods**. 3.ed. Zurich: ISTA, 1995.

LIMA-BRITO, A. *et al.* Effect of gibberellic acid (GA_3) on the emergence of plantlets of Annona crassiflora Mart., Annona squamosa L. and Annona muricata L. **Magistra**, Cruz das Almas-BA, v. 18, n. 1, p. 27-33, 2006.

MARCOS FILHO, J. **Physiology of seeds of cultivated plants.** Piracicaba: FEALQ, 2005.

METIVIER, J. R. **Cytokinins and gibberellins**. In: FERRI, M. G. Fisiologia vegetal. 2ed. Sao Paulo: EDUSP, v. 2, 1986.

MORAIS, O. M. *et al.* **Storage of seeds of** *Anonna squamosa* L**.** Biotemas (UFSC), v. 22, p. 33-44, 2009.

NAKAGAWA, J. Vigor tests based on plant performance. In: KRZYZANOWSKI, F.C.; VIEIRA, R.D.; FRANQA NETO, J.B. (Ed.). Seed vigor: concepts and tests. Londrina: **ABRATES**, Chap. 2, p.1-24, 1999.

PEREIRA, E. B. C. *et al.* **Breaking the Dormancy of Araticum Seeds**. Planaltina, DF: Embrapa Cerrados, Research and Development Bulletin, No. 137, 2004.

PINA-RODRIGUES, F.C.M.; JESUS, R.M. Contribuigao para a definigao de padrões de germinagao das especies arboreas da Mata Atlantica. II. *Poeppigia procera* (Presl.) and *Clarisia racemosa* Ruiz & Pav. In: **Anais** do VIII Congresso da Sociedade Botanica de Sao Paulo,Campinas, SP,

September 9-15, 1990a.

POPINIGIS, F. **Fisiologia da semente**. 2. ed. Brasilia, DF: [s.n.], 1985. 289 p.

RANIERI, B. D. *et al.* Seed germination of *Lavoisiera cordata* Cogn. and *Lavoisiera francavillana Cogn.* (Melastomataceae), sympatric species from Serra do Cipo. Brazil. **Acta Botanica Brasilica.** vol.17 n.4 Sao Paulo Oct./Dec. 2003.

RIBEIRO, J. F.; WALTER, B. M. T. **Phytophysiognomies of the Cerrado Biome**. In Cerrado: Environment and Flora. Embrapa/CPAC, Planaltina, 1998. p. 87-166.

RIBEIRO, J. F. *et al.* **Araticum** *(Annona crassiflora* MART.). Jaboticabal: Funep, 2000. 52p.

RIZZINI, C. T. & MORS, W. B. **Botanica economica brasileira**. Sao Paulo, EPUIEDUSP. 1976.

SALOMAO, A. N. *et al.* **Germination of seeds and seedling production of cerrado plants**. Brasilia: Cerrado Seed Network, 2003. 96p.

STENZEL, N. M. C.; MURATA, I. M.; NEVES, C. S. V. J. Overcoming dormancy in Atemoia and Fruta-do-Conde seeds. **Revista Brasileira de Fruticultura**, v. 25, n. 2, p. 305-308, 2003.

SILVA, J. A.; SILVA, D. B.; JUNQUEIRA, N. J.; ANDRADE, L. R. M. **Frutas Nativas dos Cerrados**. Empresa Brasileira de Pesquisa Agropecuaria do Brasil (EMBRAPA), Brazil, p. 50-149, 1994.

SOUZA NETO, R. A. de; CARVALHO, J. E. U. de; MULLER, C. H. Germination of araticum-do-brejo seeds *(Annona glabra* L.) submitted to pre-soaking in gibberellic acid. In Brazilian Congress of

Fruticultura, 17, 2002, Belem, PA. **Proceedings.** Belem: SBF, 2002. 4f. 2005.

TORRES, S. B. *et al.* Electrical conductivity test on sesame seeds. **Revista Brasileira de Sementes**, vol. 31, n° 3, p.070-077, 2009.

VIEIRA, R.D.; KRZYZANOWSKI, F.C. Electrical conductivity test. In: KRZYZANOWSKI, F.C.; VIEIRA, R.D.; FRANQA NETO, J.B. (Eds.). **Seed vigor**: **concepts and tests.** Londrina: ABRATES, p.4-1 to 4-26, 1999.

CHAPTER II

BEHAVIOR OF GRAVIOLA SEEDS BY GA_3 ACTION IN DIFFERENT SUBSTRATES

BEHAVIOR OF GRAVIOLA SEEDS BY GA3 ACTION IN DIFFERENT SUBSTRATES

ABSTRACT: Soursop is a fruit tree of growing interest in Brazil due to its sweet fruit. The aim of this study was to evaluate the behavior of soursop seeds by applying gibberellic acid (GA3) in different substrates. After collection, extraction and asepsis procedures, the seeds were soaked for 12 hours in their respective treatments with different concentrations of GA_3 . To assess the germination behavior of the seeds, 4 treatments (T) were carried out, with the following concentrations of GA_3 : T1 (control) - 0 mg, T2 - 500 mg, T3 - 1000 mg; T4 - 2000 mg. Four different substrates were used for each treatment: sand and vegetable soil, sand, cerrado soil and germitest paper. The parameters evaluated were germination percentage and germination speed index. From the analysis of the results, it can be concluded that the soursop seeds behaved as recalcitrant and that the best methodology for producing soursop seedlings is to subject the seeds to the following combination: T2 and the use of germitest paper in a controlled chamber; however, the substrate sand with vegetable soil in a greenhouse also offers excellent results.

KEYWORDS: Graviola; phytoregulator; seed physiology.

INTRODUCTION

The soursop, *Annona muricata* L, is a fruit tree from the Annonaceae family, native to Central America. Its sweet-tasting fruits are eaten *fresh and* used in the food, pharmaceutical and cosmetics industries.

In Brazil, it is grown mainly in the northeastern states, but due to its adaptability, it has been cultivated by small producers in almost all regions of the country. This can be seen in the increase in the supply of the fruit at street markets. Graviola is already one of Brazil's tropical fruit trees, with excellent prospects for commercial exploitation. The prospects for anonaceous cultivation are due to the high prices achieved in the market, and their inclusion in the European and American markets (SOBRINHO, 2010).

In addition to the fruit, other parts of the soursop tree are used in medicine for their diuretic, antispasmodic, antiscorbutic and astringent properties, as well as being a source of calcium. The search for new treatments and new substances to help in the treatment of cancer, existing substances in soursop have been studied to combat some types of cancer (SILVA and NEPOMUCENO, 2011).

The soursop tree reaches an average height of 4 to 8 meters in adulthood, with a single stem and asymmetrical branching. Fruiting begins between the third and fifth year, with full development between the sixth and eighth year (FREITAS, 1997). The leaves have a short petiole, are oblong-lanceolate or elliptical, 14 to 16 cm long and 5 to 7 cm wide; the veins are barely perceptible (MANICA, 1997).

The flowers in the capitulum stage have a subglobose or pyramidal shape, are hermaphrodite, dark green when growing and light green when close to anthesis, distributed on short axillary peduncles or directly on the stem, solitary or in groups of 2 to 4 flowers (MANICA, 1997). The calyx is made up of three small sepals and the corolla of six fleshy petals, formed by two whorls, the outer of which is pre-flowering valvate. The androecium is made up of several stamens, clavate and crowded, with short fillets. Each has two anthers which open lengthways to release the pollen. The gynoecium is made up of numerous pistils grouped in the shape of vaults above the stamens; the ovary is superoecious, unilocular and uniovulate; and the stigma is short with a single stigma (PINTO e SILVA, 1994). The fruit is a compound berry, multiple fruit or syncarp, fleshy, weighing from 0.4 to 10 kg. The pulp is white, juicy and slightly acidic, very aromatic and high in retinol (vitamin A) and ascorbic acid (vitamin C) (MANICA, 1997). The seeds measure between 1 and 2 cm in length, have an average weight of 0.59g and are usually black when removed from the fruit, turning brown a few days later. They usually number more than 100 per fruit (PINTO and SILVA, 1994).

The seeds have germination-inhibiting substances that cause dormancy which, together with the resistant and impermeable tegument, provide factors that are antagonistic to rapid and uniform germination (PAWSHE *et al.,* 1997). In addition to dormancy, reports by Bosco and Aguiar Filho (1995) show that germination of graviola seeds is slow and late.

Germination factors such as the impermeability and dormancy of seeds naturally protect them and provide conditions for survival during periods that are unsuitable for their establishment and development. The sensitivity of seeds allows them to awaken automatically when environmental conditions become suitable. Popinigis (1985) concluded that this process evolved as a mechanism for surviving adverse conditions, but that it hinders germination.

Studies with plants of the *Annona* genus have demonstrated the efficiency of using gibberellins to overcome the dormancy characteristic of this genus.

The use of gibberellic acid stimulates enzymatic activity in germination, contributes to overcoming dormancy, and together with light and low temperatures favors cell expansion, thus promoting germination (LIMA-BRITO *et al.,* 2006). According to Metivier (1986), gibberellins also act to control the hydrolysis of reserves.

The aim of this study was to evaluate the behavior of graviola seeds using GA3, by means of germination percentage (PG) and germination speed index (IVG), in different substrates and under field and laboratory environmental conditions.

METHODOLOGY

The seeds of *Annona muricata* L. were collected from mother trees in the rural area of the municipality of Ituiutaba-MG, in June 2016. The collected seeds were manually extracted from the fruit, washed and placed on paper towels to dry in the open air, in the shade, for 24 hours. Before starting the experiments, the seeds were aseptically soaked in 2% sodium hypochlorite for 10 minutes. They were washed in distilled water and soaked for 24 hours in their respective treatments containing different concentrations of GA3. The seeds were rinsed again in distilled water and distributed in seedbeds containing substrates and on germitest paper, moistened daily throughout the experiment.

To assess the germination of graviola seeds, eight treatments were carried out with the following concentrations of gibberellic acid (GA_3): treatment 1, control 0 $mg.L^{-1}$; treatment 2, 500 $mg.L^{-}$ 1; treatment 3, 1000 $mg.L^{-1}$; treatment 4, 2000 $mg.L^{-1}$, each diluted in 1L of distilled water. Four different substrates were used for each treatment: sand and vegetable soil, sand, cerrado soil and germitest paper.

The seed germination experiments on the substrates were conducted in accordance with Brasil (2009), outdoors, in natural environmental conditions in the city of Ituiutaba-MG, located in a region of the cerrado biome. The germination test on germ paper was conducted in a controlled germination chamber at 25°C. The number of germinated seeds was assessed daily and the criterion for emergence was the exposure of part of the plant to the surface of the substrate. Electrical conductivity, germination speed index (GVI) and germination percentage were also assessed.

The germination percentage was assessed for 50 days. While the GVI was assessed 10 days after germination. For each treatment, the GVI was calculated using the method indicated by Biruel *et al.* (2007), which shows the number of seeds germinated per day, with GVI = G1/N1+ G2/N2 +... Gn/Nn, where G1, G2... Gn equals the number of seeds germinated, and N1, N2... Nn corresponds to the number of days.

The electrical conductivity test was carried out according to HAMPTON; TEKRONY (1995) and VIEIRA; KRZYZANOWSKI (1999), for each treatment. The seeds were soaked in plastic cups (200 mL) containing 75mL of GA solution$_3$ from each treatment and kept in controlled conditions at 25°C for 24 hours^{-1} . After soaking, the electrical conductivity was read using a DIGIMED conductivity meter, model CD 21, with a constant 1.0 electrode and the results expressed in "micro Siemens per centimeter per gram" (pS cm g).$^{-1-1}$

The results were analyzed using a completely randomized design, with five replicates per treatment. Statistical measures of dispersion and analysis of variance (ANOVA) were calculated, applied to the factorial design and the means were compared using Tukey's test at 5% probability.

RESULTS AND DISCUSSION

Table 1 shows the average percentage moisture content of *Annona muricata* L. seeds subjected to the standard oven water content test (105°C/24h).

Table 1. Average values of the water content (TA) of *Annona muricata* L. seeds.

TREATMENT	WATER CONTENT (%)
T1	27,866 (±0,975) A
T2	27,448 (±1,251) A
T3	27,660 (±0,430) A

T4	25,768 (±0,633) A
C.V (%)	6,75

Means of five repetitions. Values followed by the same letter in the column do not differ by the Tukey test ($p < 0.05$). Values in brackets correspond to the standard deviation. CV = coefficient of variation.

The water content test showed that the batches used in the experiment had a moisture content of between 25.768% and 27.866%, and there was no significant difference between the batches used for each treatment. In this experiment, it can be said that the seeds used were viable and had a considerable water content, since in a study by Figueiredo (2012), graviola seeds withstand desiccation, maintaining the viability of the seeds with a water content of around 7%, revealing the orthodox nature of the species.

Table 2 shows the results of the electrical conductivity test of *Annona muricata* L. seeds, which were soaked for 24 hours in different treatments.

Table 2. Average values of the electrical conductivity (EC) test of *Annona muricata* L. seeds, given in pS cm^{-1} g^{-1} .

TREATMENT	ELECTRICAL CONDUCTIVITY - CE (pБ.стЛд'1)
T1	40,208 (±0,304) C
T2	73,61 (±0,235) B
T3	74,48 (±0,474) B
T4	140,88 (±0,491) A
C.V (%)	11,15

Means of five repetitions. Values followed by the same letter in the column do not differ by the Tukey test ($p < 0.05$). Values in brackets correspond to the standard deviation. CV = coefficient of variation.

In the electrical conductivity (EC) test, electrolyte leaching occurred in ascending order between treatments, with T1 showing the lowest value. Treatment T4, with the lowest vigor, differed statistically from the others (Table 2), suggesting that GA_3 at a dose of 2000mg negatively influenced the reorganization of cell membranes and led to a greater loss of exudates.

According to Rech *et al.* (1999), high electrical conductivity values are related to lower seed vigor, as they represent the leakage of solutes from cell membranes. On the other hand, excessively low values may indicate little or no physiological activity in the seeds. This can be seen when comparing the results of the electrical conductivity test with the results of the germination test, where the control with low electrical conductivity showed the lowest germination rate, and in treatment 4, which had a high electrical conductivity value, there was no germination. Treatments 2 and 3 had average electrical conductivity values and also the best germination values. These results are similar

to those found in various studies such as: MCDONALD; WILSON (1979), POWELL (1986), MARCOS FILHO *et al.,* (1990), VIEIRA *et al.,* (2002) VIEIRA *et al.,* (2004), VIEIRA (1994), VIEIRA; KRZYZANOWSKI (1999); PANOBIANCO; MARCOS FILHO (2001).

Table 3. Average germination values (%) and germination speed index (GVI) of *Annona muricata* L. seeds.

TREATMENTS	SUBSTRATE	GERMINATION (%)	IVG
T1	Germitest paper	28 (±0,48)a	4,78 (±0,37)a
	Sand+Earth	16 (±0,40)a	2,08 (±0,40)a
	Vegetable		
	Sand	14 (±0,24)a	2,83 (±0,24)a
	Land of the Cerrado	26 (±0,51)a	3,78 (±0,74)a
T2	Germitest paper	76 (±0,67)b	8,58 (±0,74)b
	Sand+Earth	52 (±0,63)b	7,24 (±0,70)a
	Vegetable		
	Sand	18 (±0,37)a	3,08 (±0,22)a
	Land of the Cerrado	20 (±0,54)a	3,58 (±0,20)a

T3	Germitest paper	34 (±0,60)a	4,53 (±0,24)a
	Sand+Earth Vegetable	29 (±0,24)a	2,83 (±0,26)a
	Sand	8,0 (±0,36)a	1,5 (±0,63)a
	Land of the Cerrado	18 (±0,37)a	3,08 (±0,22)a
C.V (%)		6,63	7,13

Averages of five replicates. Values followed by the same letter in the column do not differ according to the Tukey test ($p < 0.05$). Values in brackets correspond to the standard deviation. CV = coefficient of variation.

The data shown in Table 3 refer to the average results of the germination test (%) and the germination speed index (GVI) of *Annona muricata* L. seeds, submitted to treatments with GA_3 , in different substrates.

T1 showed insignificant germination percentages in the Tukey test ($p < 0.05$), a fact corroborated by Bosco; Aguiar Filho (1995), who found that soaking soursop seeds in water did not significantly influence germination. T2 had the best germination percentage, 76% on germ paper and 52% in sand with vegetable soil, which can be considered a good result, since Pinto (1975), in field notes, found that graviola seeds have a germination power of no more than 48%. T2 also showed the best GVI, which differed statistically from the other treatments.

Among the substrates analyzed, the best performance was seen in germ paper, both in PG and IVG, which can be explained by the controlled germination conditions, since, according to Marcos Filho (2005), four factors are basic for promoting germination: temperature, water availability, phytohormone action and the presence of light. The use of germitest paper in a controlled germination chamber made it possible to control temperature, humidity and the presence of light.

Sand with topsoil showed a high percentage of germination at T2, reaching 76%. Oliveira *et al.* (1989) corroborate this finding.
observation, as they state that high temperatures associated with substrates

with high moisture retention contribute to good results in recalcitrant seeds.

In summary, the use of gibberellic acid to break dormancy in *Annona muricata* L. seeds was efficient at a concentration of 500 mg of GA_3, providing a significant increase in the germination rate, which was even higher when the seeds were placed on germ paper and in a controlled chamber at 25°.

CONCLUSIONS

It was concluded that soursop seeds behaved as recalcitrant and that the best method for producing soursop seedlings is to subject the seeds to a combined treatment of 500 mg of GA_3 and to use germitest paper as a substrate in a controlled chamber. It is worth noting that the substrate sand with vegetable soil, in the greenhouse, also showed good results, as well as being more accessible.

REFERENCES

ARAUJO, J. F. **Treatments to accelerate and standardize the germination of pine cone seeds *(Annona squamosa* L.).** Dissertation (Master's Degree in Agricultural Sciences) Federal University of Bahia, Cruz das Almas. 1991, 56f.

BARBOSA, Z.; SOARES, I.; CRISOSTOMO, L. A. Growth and nutrient uptake by soursop seedlings. **Revista Brasileira de Fruticultura**, Jaboticabal - SP, v. 25, n. 3, p. 519-522, 2003.

BIRUEL R. P; AGUIAR I. B; PAULA R. C. Germination of pau- ferro seeds submitted to different storage conditions, chemical scarification, temperature and light. **Revista Brasileira de Sementes**, v.29, n.3, p.134-141, 2007.

BOSCO, J.; AGULAR FILHO, S. P. de. Overcoming dormancy in graviola seed *Annona muricata* L. **ABRATES**. Informativo Abrates, Londrina, v.5, n.2, p.93, 1995.

BRAZIL. Ministry of Agriculture. **Rules for Seed Analysis**. Brasilia: SNAD/DNPV/CLAV, 1992. 365p.

FIGLIOLIA M. B.; OLIVEIRA E. C.; Pina-Rodrigues F. C. M. Seed analysis. In: Aguiar IB, Pina-Rodrigues F. C. M, Figliolia M. B., coordinators. Tropical forest seeds. Brasilia: **ABRATES**, 1993.

FIGUEIREDO, G. R. G. Propagation of soursop: seed vigor under desiccation, cuttings and mini-stakes. 2012. 67f. Dissertation (Master's in Plant Production) - University of Santa Cruz, Ilheus, 2012.

FREITAS, G. B. **Propagation, flowering, fruiting and production of the soursop tree** *(Annona muricata* L.). Viposa: UFV, 1987. 87p. Thesis (Doctorate in Plant Science) - Federal University of Viposa, 1997.

GOUVEIA, D. S. *et al.* Physico-chemical evaluation and sensory acceptance test of pine cone juice and pine cone-milk blend. **Revista Brasileira de Produtos Agroindustriais**, Campina Grande, v.9, n.1, p.29-36, 2007.

HERNANDEZ, L.V. **La reproduccion sexual y multiplicacion vegetativa de la annonaceas**. Xalapa: Universidad Veracruzana, 35p. 1993.

LEDERMAN, I. E.; BEZERRA, J. E. F. Current situation and prospects for anonaceas in the state of Pernambuco. In: SAO JOSE, A R. *et al.* (eds.) **Anonaceas:** Tecnologia de produpao e comercializapao, Vitoria da Conquista, BA: DFZ/UESB, p. 173-177, 1997.

LIMA-BRITO, A. *et al.* Effect of Giberelic Acid (GA_3) on the emergence of plantlets of *Annona crassiflora* Mart.,*Annonasquamosa* L. and *Annonamuricata* L. **Magistra,** Cruz das Almas, V. 18, n. 1, p. 27-33, 2006.

MANICA, I. Taxonomy, morphology and anatomy. In: SAO JOSE, A. R.; SOUZA, I. V. B.; MORAIS, O. M.; REBOUQAS, T. N. H. **Anonaceas, Produpao e Mercado**: Pinha, Graviola, Atemoia e Cherimolia. Vitoria da Conquista: DFZ/UESB, p.20- 35, 1997.

MARCOS FILHO, J.; Cicero, S. M.; Silva, W. R. **Avaliapao da qualidade das sementes**. Piracicaba: FEALQ, 230 p., 1987.

McDONALD Jr., M. B.; WILSON, D. O. An assessment of the standardization and abiliity of the ASA - 610 to rapidly predict potential and soybean germination. **Journal of Seed Technology**, Springfield, v.4, n.1, p.1-11, 1979.

MELO, G. S.; NETO, L. G.; MOURA, R. J. M. **Cultivation of the soursop tree** *(Annona muricata* L.). Recife, PE. 1983. 4 p. (Instrupoes Tecnicas do IPA; 13).

MENEGAZZO, M. L. et al. Effects of methods to overcome dormancy in pine cone seeds *(Annona squamosa* L.). **Agrarian Journal**, v.5, n.15, p.29-35, 2012.

METIVIER, J. R. **Cytokinins and gibberellins**. In: FERRI, M. G. Fisiologia vegetal. 2ed. Sao Paulo: EDUSP, 1986. v.2, p.93-162.

OLIVEIRA, E. C.; PINA-RODRIGUES, F. C.; FIGLIOLIA, M. B. Proposals for the standardization of methodologies in forest seed analysis. **Revista Brasileira de Sementes**, Brasilia, v.11, n. 1-3, p. 1-42, 1996.

PANOBIANCO M.; MARCOS FILHO J. Accelerated aging and controlled deterioration in tomato seeds. **Scientia Agricola** 58: p.525-531, 2001.

PAWSHE, Y. H.; PATIL, B. N.; PATIL, L. P. Effect of pre-germination seed treatment on the germination and vigour of seedlings in custard apple *(Annona squamosa* L.). **Annals of Plant Physiology**, v. 11, n. 2, p. 150-154, 1997.

PINTO, A. C. de Q.; SILVA, E. M. da. **Graviola for export**: technical aspects of production. Brasilia: EMBRAPA, SPI (Serie Publicapoes Tecnicas FRUPEX, 7), 41p. 1994.

POPINIGIS, F. **Fisiologia da semente**. 2ed. Brasilia, DF: [s.n.], 1985. 289 p.

POWELL, A. A. Cell membranes and seed leachate conductivity in relation to the quality of seed sowing. **Journal of Seed Technolology,** v.10, n.2, p.81100, 1986.

RECH E. G., VILLELA, F. A., TILLMANN, M. A. Rapid evaluation of the physiological quality of pea seeds. Revista Brasileira de Sementes 1999; 21(2): 1-9.

SACRAMENTO, C. K.; MOURA, J. I. L.; COELHO JUNIOR, E. Graviola. In: SANTOS-SEREJO, J. A. *et al.* (eds.). **Tropical fruit growing**: regional and exotic species. Brasilia, DF: Embrapa Informagao Tecnologica, p.95132, 2009.

SILVA, L. M.; NEPOMUCENO, J. C. Modulatory effect of graviola pulp *(Annona muricata)* on the carcinogenicity of mitomycin C, evaluated using the *warts* test in *Drosophila melanogaster.* **PERQUIRERE**, Patos de Minas: UNIPAM, n. 8, vol. 1, p. 80-94, jul.2011.

SOBRINHO, R. B. **Potential for the exploitation of anonaceae in the Northeast of Brazil.** 17ª Semana da Fruticultura, Floricultura e Agroindustria. Embrapa Agroindustria Tropical. Fortaleza-CE, 2010.

STENZEL, N. M. C.; MURATA, I. M.; NEVES, C. S. V. J. Overcoming dormancy in seeds of atemoia and fruta-do-conde. **Revista Brasileira de Fruticultura**, Jaboticabal - SP, v. 25, n. 2, p. 305-308, agosto 2003.

VIEIRA, R. D.; CARVALHO, N. M. Editors. **Seed vigor tests**. Jaboticabal: FUNEP; 1994.

VIEIRA, R. D. *et al.* Electrical conductivity and initial water content of soybean seeds. **Pesquisa Agropecuaria Brasileira**, v.37, n.19, p.13331338, 2002.

VIEIRA, R. D. *et al.* Electrical conductivity of the seed soaking solution and soybean seedling emergence. **Scientia Agricola**, v.61, n.2, p.164-168, 2004.

VIEIRA, R. D.; KRZYZANOWSKI, F. C. Electrical conductivity test. In: KRZYZANOWSKI, F. C.; VIEIRA, R. D.; FRANQA NETO, J. B. (Eds.). **Seed vigor**: concepts and tests. Londrina: ABRATES, p.4-1 to 4-26, 1999.

CHAPTER III

EFFECT OF GA3 ON THE PHYSIOLOGICAL QUALITY OF PINE SEEDS IN DIFFERENT SUBSTRATES

EFFECT OF GA3 ON THE PHYSIOLOGICAL QUALITY OF PINE SEEDS IN DIFFERENT SUBSTRATES

ABSTRACT: The dormancy of pine cone seeds, *Annona squamosa* L., makes it difficult to produce seedlings. The aim of this study was to evaluate the physiological quality of pine cone seeds treated with different concentrations of gibberellic acid (GA_3) and in different substrates. After manual extraction and asepsis procedures, the seeds were soaked for 12 hours in the following treatments (T), with the following concentrations of gibberellic acid (GA_3): T1 (control) - 0 mg L^{-1} , T2- 200 mg L^{-1} , T3- 400 mg L^{-1} ; T4- 800 mg L^{-1} . After soaking, the seeds were placed in seedbeds containing the following substrates: sand and vegetable soil, sand, and cerrado soil. The experiments were carried out in a greenhouse. The parameters evaluated were germination percentage over 60 days and germination speed index over 30 days. The data obtained was compared using the Tukey test at 5% probability. From the analysis of the results, it can be concluded that GA_3 applied in different concentrations to pine cone seeds had no protective effect on cell membranes; however, it proved to be efficient in the germination process at concentrations of 200, 400 and 800 mg L^{-1} , and when the seeds were sown in a sand substrate with vegetable soil, the results were maximized.

KEYWORDS: Pine cone; phytoregulator; seed physiology.

INTRODUCTION

Annona squamosa L. belongs to the Annonaceae family and is a small fruit tree; its fruit is popularly known as pine cone, ata or fruta-do-conde, it has soft flesh and a pleasant flavor.

The pine tree originated in the Antilles and is widely distributed in tropical and subtropical regions of the world. In Brazil, it is grown mainly in the northeastern states (ARAUJO *et al.,* 1999), which invest resources in irrigation and generate employment and income for the people of the semiarid region (MENEZES *et al.,* 2002), a region where the most important crops are the pine trees.
climatic aspects favor fruit production during all months of the year (DIAS *et al.,* 2004).

The pine cone stands out among the cherimoya species because it is the most commercial (MENEGAZZO *et al.,* 2012), used in human food *in natura* (pulp) because it is rich in carbohydrates that act as excellent sources of energy and mineral salts such as calcium, phosphorus, potassium; important in various biological functions such as: protecting bones, teeth, as well as making muscles firmer (PINTO *et al.,* 2005).

The intense production of seedlings for the formation of new orchards is precisely the result of the growing interest in this fruit. To produce anonaceous seedlings, grafting is the most suitable form of propagation, and the rootstock must be obtained from seeds (GEORGE; & NISSEN, 1987; GAMA & MANICA, 1994).

The pine tree is basically reproduced by seeds; however, the crop can be propagated vegetatively to select more productive clones (FERREIRA *et al.,* 2002).

Pine seeds are difficult to germinate due to the presence of inhibitory substances, as well as the fact that their seed coat is resistant and impermeable (MENEGAZZO *et al.* 2012).

Seed germination is a complex process that develops through a series of metabolic and morphogenetic events that transform the embryo into a plant (CASTRO *et al.,* 2005), however, this process in pine cones is hampered by the existence of dormancy in their seeds.

There are two types of dormancy: morphological, exogenous or tegumentary, resulting from the impermeability of the tegument to water or gases, and which can be overcome through post-maturation processes of the embryo; and physiological, endogenous or embryonic, resulting from the immaturity of the embryo or inhibition of germination. To overcome the latter, it is necessary to add hormones and phyto-regulators, wash the seeds for certain periods, carry out heat treatment and other methods (FOWLER; BIANCHETTI, 2000).

Pine seeds have no physical impediment to water ingress (LEMOS *et al.,* 1988). Ferreira *et al.* (1997) state that seed dormancy in this species is not related to the hardness of the tegument, but rather to the dormancy mechanism for seasonal survival.

For nurserymen and producers, the dormancy mechanism is a disadvantage because it induces great unevenness between seedlings and a greater demand for time in their production (EIRA *et al.,* 1993). Therefore, the study of methodologies to improve seed germination and seedling performance in the nursery is important and helps to speed up and standardize the initial establishment of plants in the field (ROVERSI *et al.,* 2002).

Accelerating and standardizing the germination process in pine cone seeds has been achieved by applying some treatments (STENZEL *et al.,* 2003), but with inconsistent results and variations in recommendations.

Dormancy can also result from a hormonal balance between growth promoters and inhibitors (WEAVER, 1987).

Gibberellin, an endogenous growth regulator, plays an important role in plant growth and development, especially in inducing seed germination, as well as promoting hypocotyl and stem elongation, regulating pollen development and inducing flowering (PENG; HARBERD, 2000 and RICHARDS, 2001).

Menegazzo *et al.* (2012) say that the best method for overcoming the dormancy of *Annona squamosa* L. seeds is to use gibberellic acid, although the best concentration of this phyto-regulator has yet to be defined by new experiments.

The action of gibberellic acid (GA_3) on seed germination has been investigated by authors such as Macchia *et al.* (2001) on *Echinacea angustifolia* DC (Asteraceae), Passos *et al.,* (2004), on *Passiflora nitida* KUNTH, for its use in *Annona squamosa* L., aiming to break dormancy in seeds.

The aim of this study was to evaluate the physiological quality of pine cone seeds subjected to different treatments with gibberellic acid (GA3) in different substrates.

METHODOLOGY

The seeds of *Annona squamosa* L. were collected from mother trees in the rural area of the municipality of Ituiutaba-MG, in February 2017. They were extracted manually from the fruit, washed and placed on paper towels to dry in the open air, in the shade, for 24 hours. They were then aseptically treated in 2% sodium hypochlorite for 10 minutes, washed in distilled water and soaked in their treatments with different concentrations of GA_3 for 24 hours. The seeds were washed again in distilled water and placed in seedbeds containing the substrates and on germ paper, moistened daily throughout the experiment.

To assess the germination of pine cone seeds, eight treatments (T) were carried out with the following concentrations of gibberellic acid (GA_3): T1, control - 0 $mg.L^{-1}$; T2 - 200 $mg.L^{-1}$; T3 - 400 $mg.L^{-1}$; T4 - 800 $mg.L^{-1}$.

Four different substrates were used for each treatment: sand and topsoil, sand, cerrado soil and germitest paper.

The seed germination experiments on the substrates were conducted in accordance with Brasil (2009), outdoors, in natural environmental conditions in the city of Ituiutaba-MG, located in a region of the cerrado biome. The germination test on germ paper was conducted in a controlled germination chamber at 25°C. The number of germinated seeds was assessed daily, with the criterion for emergence being the exposure of part of the seedling on the surface of the substrate. The Germination Speed Index and Germination Percentage were also evaluated.

Germination percentage was assessed for 50 days. The Germination Speed Index was assessed 10 days after germination. For each treatment, the Germination Speed Index (GVI) was calculated, as suggested by Biruel *et al.* (2007), which shows the number of germinations.

germinated seeds per day, where GVI = G1/N1+ G2/N2 +... Gn/Nn, where G1, G2... Gn equals the number of seeds germinated, and N1, N2... Nn corresponds to the number of days.

The Electrical Conductivity (EC) test was carried out according to HAMPTON; TEKRONY (1995) and VIEIRA; KRZYZANOWSKI (1999), for each treatment. The seeds were soaked in 200 mL plastic cups containing 75 mL of the GA3 solution from each treatment and kept in controlled conditions at 25°C for 24 $hours^{-1}$. After soaking, the electrical conductivity was read using a DIGIMED conductivity meter, model CD 21, with a constant 1.0 electrode. The results were expressed in "micro Siemens per centimeter per gram" (pS cm g $)^{.-1-1}$

The results were analyzed using a completely randomized design, with five replications per treatment. Statistically, the measures of dispersion and analysis of variance (ANOVA) were calculated, applied to the factorial scheme and the means were compared using Tukey's test at 5% probability.

RESULTS AND DISCUSSION

Table 1 shows the results of the standard water content test carried out in an oven (105°C/24h).

Table 1. Average values of the water content test of *Annona squamosa* L. seeds.

TREATMENT	WATER CONTENT (%)
T1	24,755 (±0,975) A
T2	24,337 (±1,251) A
T3	24,550 (±0,430) A
T4	24,657 (±0,633) A
C.V (%)	4,95

Means of five repetitions. Values followed by the same letter in the column do not differ by the Tukey test ($p < 0.05$). Values in brackets correspond to the standard deviation. CV = coefficient of variation.

The water content of the seeds is related to the relative humidity which controls the occurrence of the different metabolic processes that the seed can undergo (CARVALHO and NAKAGAWA, 2000). It was possible to observe in the water content test that the lots used in the four treatments in the experiment had uniform moisture content, with no significant difference between them. Morais (2009), when analyzing the water content of pine cone seeds and comparing it with the germination percentage, observed an increase in the germination percentage as the water content decreased.

Table 2. Average values of the Electrical Conductivity test of *Annona squamosa* L. seeds, subjected to soaking for 24 hours with different treatments, given in pS cm^{-1} g^{-1} .

ELECTRICAL CONDUCTIVITY TREATMENT - CE (рБ.стЛд'1)	
T1	20,309 (±0,403) B
T2	51,845 (±0,354) A

T3	54,361 (±0,632) A	
T4	56,056 (±0,402) A	
C.V (%)		7,62

Averages of five replicates. Values followed by the same letter in the column do not differ according to the Tukey test ($p < 0.05$). Values in brackets correspond to the standard deviation. CV = coefficient of variation.

In the Electrical Conductivity test, electrolyte leaching occurred in ascending order between treatments, with T1 showing the lowest value and differing from the others. Although excessively low values may indicate little or no physiological activity in the seeds, Oliveira (2012) states that the lower values obtained represent less exudate release, and therefore greater seed vigor and less disorganization in the cell membrane systems. This relationship can be observed when comparing the results of the electrical conductivity test with the results of the germination test, in which the control with a low electrical conductivity value showed lower germination than treatments 2, 3 and 4, in which higher electrical conductivity values favored higher germination.

The seeds of *Annona squamosa L.* were subjected to the germination speed index and germination percentage tests, but the seeds obtained from the market did not germinate. The results shown in Table 3 refer only to the seeds collected.

Table 3. Average germination values (%) and germination speed index of *Annona squamosa* L. seeds, subjected to different treatments (T1, T2, T3 and T4) with GA_3, and in different substrates.

SUBSTRATE TREATMENTS		GERMINATION (%)	IVG
	Sand	30,0 (±0,54)b	4,23 (±0,019)a
T1	Terra of Cerrado	37,0 (±0,87)b	4,94 (±0,017)a
	Sand + Terra Vegetable	39,0 (±0,66)b	5,02 (±0,026)a

T2	Sand	58,0 (±0,50)b	6,16 (±0,032)a
	Terra of Cerrado	62,0 (±0,24)a	6,36 (±0,025)a
	Sand + Vegetable Terra	69,0 (±0,37)a	6,73 (±0,023)b
T3	Sand	63,0 (±0,60)a	6,40 (±0,026)a
	Terra of Cerrado	66,0 (±0,37)a	6,61 (±0,021)b
	Sand + Vegetable Terra	74,0 (±0,58)a	7,03 (±0,013)b
T4	Sand	59,0 (±0,37)b	6,22 (±0,030)a
	Terra of Cerrado	68,0 (±0,67)a	6,67 (±0,024)b
	Sand + Topsoil	86,0 (±0,89)a	7,08 (±0,023)b
	C.V (%)	8,80	15,02

Averages of five replicates. Values followed by the same letter in the column do not differ according to the Tukey test ($p < 0.05$). Values in brackets correspond to the deviation standard. CV coefficient of variation.

The best GVI was in T4, combined with the substrate sand with vegetable soil, with 86% seed germination. Also in T4, 68% germination was obtained in the cerrado soil substrate and 59% germination in sand.

In T3, 74% of the seeds germinated in the sand substrate with vegetable soil, followed by 66% in the cerrado soil and 63% in the sand. T2 showed 69% germination in the sand substrate with vegetable soil, 62% in the cerrado soil and 58% in sand.

With regard to the use of gibberellic acid, the best concentration was 800 mg L^{-1} . There are recommendations for the use of GA_3 at dosages of 50 mg L^{-1} (STENZEL *et al.,* 2003) to 750 mg L^{-1} (ARAUJO, 1991). Hernandez (1993) reported that GA_3 used at a concentration of 100 mg L^{-1} promoted a significant increase in the germination of *Annona cherimola* L. from 57.25% (control) to 70.00%. Valenzuela and Osorio (1998) obtained 55.4% germination in condessa seeds *(Annona reticulata)* with 10,000 ppm of GA_3.

According to Ramos *et al.* (2002), a good substrate is one that aims to provide suitable conditions for germination and/or the emergence or development of the root system of the seedling in formation. The best substrate was sand with topsoil, for all treatments, and this had the highest GVI. Oliveira *et al.* (1996) corroborate this finding.

It is worth noting that high temperatures associated with substrates with high moisture retention can produce good results for recalcitrant seeds, as is the case here, since topsoil has good water retention and sand provides high temperatures in the substrate.

The terra do cerrado substrate showed average performance in the germination of pine cone seeds, both in the Germination Percentage test and in the Germination Speed Index, which can be explained by the distribution of the pore diameters which influences the amount of water retained (HILLEL, 1980). According to Souza *et al.* (2007), there is great variation in the germination performance of seeds in relation to the substrates used in laboratory conditions, and in field conditions, cerrado soil can be a good option for germinating pine cone seeds, as it resembles the soil of the natural environment in which they grow.

The IVG was higher in the substrates terra do cerrado and sand with vegetable soil in treatments 2, 3 and 4, which differed significantly from the others. Sand had the lowest germination rate. According to Figliolia *et al.* (1993), in sand, water passes quickly into the subsurface layers. The lack of moisture may have delayed seed germination.

CONCLUSIONS

It can be concluded that GA_3 applied in different concentrations to pine cone seeds did not have a protective effect on cell membranes, however, it proved to be efficient in the germination process at concentrations of 200, 400 and 800 mg L^{-1} , and when sowing in sand substrate with vegetable soil, the results are maximized.

REFERENCES

ARAUJO, J. F. **Treatments to accelerate and standardize the germination of pine cone seeds** *(Annona squamosa L.).* Dissertation (Master's Degree in Agricultural Sciences) Federal University of Bahia, Cruz das Almas, 56f, 1991.

ARAUJO, J. F; ARAUJO J. F.; ALVES, A. A. C. **Instrupoes tecnicas para o cultivo da pinha** *(Annona squamosa* L.). Salvador: EBDA, 44p. il. (EBDA: Circular tecnica, n.7), 1999.

BIRUEL R. P; AGUIAR I. B; PAULA R. C. Germination of pau- ferro seeds submitted to different storage conditions, chemical scarification, temperature and light. **Revista Brasileira de Sementes**, v.29, n.3,
p.134-141, 2007.

BRAZIL. Ministry of Agriculture, Livestock and Supply. **Rules for Seed Analysis**. Brasilia: Mapa/ACS, 2009. 399p.

CARVALHO, N. M. de; NAKAGAWA, J. **Seeds: science, technology and production.** 4. ed. Jaboticabal: FUNEP, 2000.

CASTRO, P. R. C.; KLUGE, R. A.; PERES, L. E. P. **Manual of Plant Physiology**. Piracicaba: Editora Agronomica Ceres, 2005. P. 49.

DIAS, N. O. *et al.* Pine tree fruit growth as a function of pruned branch length. **Proceedings of** the 18th Brazilian Congress of Fruit Growing, Florianopolis, Brazil, CD-Rom, 2004.

EIRA, M. T. S.; FREITAS, R. W. A.; MELLO, C. M. C. Overcoming seed dormancy in *Enterolobium contortisiliquum* (Vell.) Morong.- Leguminosae. Revista Brasileira de Sementes, Londrina-PR, v. 15, n. 2, p. 177-181, 1993.

FERREIRA, G. *et al.* Inhibition study of sugar apple *(Annona squamosa* L.) and atemoya *(Annona squamosa* L. x A. *cherimoya* Mill.) seeds. **Proceedings of** the 1st International Congress of Annonaceae, Chapingo, Mexico, p.210-224, 1997.

FERREIRA, G.; ERIG, P. R.; MORO, E. Use of gibberellic acid in seeds of conefruit *(Annona squamosa* L.) for the production of seedlings in different packages. **Revista Brasileira de Fruticultura**, 24 (1): 178-182, 2002.

FIGLIOLIA M. B.; OLIVEIRA E. C.; PINA-RODRIGUES, F. C. M. Seed analysis. In: Aguiar IB, Pina-Rodrigues F. C. M, Figliolia M. B., coordinators. Tropical forest seeds. Brasilia: **ABRATES**; 1993.

FOWLER, A. J. P.; BIANCHETTI, A. **Dormancy in forest seeds**. Embrapa/ Embrapa Florestas, Colombo, Brazil, 27pp. (Documentos, 40), 2000.

GAMA, F.; MANICA, I. Propagation. In: MANICA, I. **Cultivation of anonaceae**:
Ata, Cherimolia, Graviola. Porto Alegre: EVANGRAF, p. 30-37, 1994.

GEORGE, A. P.; NISSEN, R. J. Propagation of Annona species, a review. **Scientia Horticulturae**, Amsterdam, v. 33, p. 75-85, 1987.

HAMPTON, J. G.; TEKRONY, B. M. (Eds.). **Handbook of vigour test methods**. 3.ed. Zurich: ISTA, 1995.

HERNANDEZ, L. V. **Sexual reproduction and vegetative multiplication of annonaceae**. Xalapa: Universidad Veracruzana, 1993. 35p.

HILLEL, D. **Fundamentals of soil physics**. New York: Academic, 1980.

LEMOS, E. E. P *et al.* Germination of pine cone seeds submitted to treatments to break dormancy. In: CONGRESSO BRASILEIRO DE FRUTICULTURA, 9., 1987, Campinas. **Proceedings**... Campinas: SBF, v.2, p.675678. 1988.

MACCHIA, M.; ANGELINI, L. G.; CECCARINI, L. Methods to overcome seed dormancy in *Echinacea angustifolia* DC. **Scientia Horticulturae**, St. Louis, v. 89, n.4, p.317-324, 2001.

MENEGAZZO, M. L. *et al.* Effects of methods for overcoming dormancy in pine cone seeds *(Annona squamosa* L.). **Agrarian Journal**, v.5, n.15, p.29-35, 2012.

MENEZES, L. S. *et al.* Effect of substrates on the production of pine seedlings *(Annona squamosa* L.) in Styrofoam trays. ***Proceedings of*** the 17th Brazilian Congress of Fruit Growing, Belem, Brazil, CD-Rom, 2002.

MORAIS, O. M. *et al.* Storage of *Anonna squamosa* L. seeds. **Biotemas** (UFSC) , v. 22, p. 33-44, 2009.

OLIVEIRA, E. C.; PINA-RODRIGUES, F. C.; FIGLIOLIA, M. B. Proposals for the standardization of methodologies in forest seed analysis. **Revista Brasileira de Sementes**, Brasilia, v.11, n. 1-3, p. 1-42, 1996.

OLIVEIRA, S. **Forest seed technology**: native species.
Curitiba: UFPR; 2012.

PASSOS *et al.* Use of gibberellic acid to break dormancy in *Passiflora nitida* KUNTH seeds germinated in vitro. **Revista Brasileira de Fruticultura**, Jaboticabal - SP, v. 26, n. 2, p. 380-381, August 2004.

PENG, J.; HARBERD, N. P. The role of GA-mediated signalling in the control of seed germination. **Current Opinion in Plant Biology**, London, v.5, p.376381, 2000.

PINTO, A. C. de.Q. *et al.* **Annonas Species**. Southampton: ICUC, 2005. 268p.

RAMOS, J. D. et al. Seedling production of fruit plants. **Informe Agropecuario**, Belo Horizonte, v. 23, n. 216, 2002.

RICHARDS, D.E. *et al.* How gibberellin regulates plant growth and development: a molecular genetic analysis of gibberellin signaling. **Annual Review of Plant Physiology and Plant Molecular Biology**, Palo Alto, v.52, p.67-88, 2001.

ROVERSI, T. *et al.* Overcoming dormancy in black acacia *(Acacia mearnsii* Willd.) seeds. **Revista Brasileira de Agrociencia**, v.8, n.2, p.161163, 2002.

SOUZA E. B. *et al.* Germination of *Adenanthera Pavonina* L. seeds as a function of different temperatures and substrates. **Revista Arvore**, 2007; 31(3):437-443.

STENZEL, N. M. C.; MURATA, I. M.; NEVES, C. S. V. J. Overcoming dormancy in seeds of atemoia and fruta-do-conde. **Revista Brasileira de Fruticultura**, Jaboticabal - SP, v. 25, n. 2, p. 305-308, August 2003.

VALENZUELA, J. R. C.; OSORIO, J. D. B. Efecto del acido giberelico y el metodo de siembra en la germinacion de semillas y crecimiento de plantulas de anona colorada *(Annona reticulata* L.). **Revista Facultad Nacional de Agronomia**, Medellin, v.51, n.2, p.235-244, 1998.

VIEIRA, R. D.; KRZYZANOWSKI, F. C. Electrical conductivity test. In: KRZYZANOWSKI, F. C.; VIEIRA, R. D.; FRANQA NETO, J. B. (Eds.). **Seed vigor: concepts and tests.** Londrina: ABRATES, 1999. p.4-1 to 4-26.

WEAVER, R.J. **Reguladores del crecimiento de las plantas en la agricultura**. 5. ed. Mexico: Trillas, 1987. 622p.

FINAL CONSIDERATIONS

The driving force behind this work was the importance of the flora species of the Cerrado biome, especially those of the *Annona* genus, and the urgent need to create reforestation measures with native species that are also economically important, as a way of mitigating the high rate of deforestation,

Knowledge about the physiology and understanding of how phytoregulatory substances break dormancy in *Annona* seeds and stimulate germination is scarce. Gibberellic acid, GA_3 , was selected to evaluate its influence on the germination of seeds of the species *Annona cacans* Warm., A. *muricata* L. and A. *squamosa* L., and concomitantly, to evaluate the efficiency of different substrates in this process.

Given the importance of vigor tests in determining the physiological quality of seeds, the following aspects were observed: water content, electrical conductivity, germination speed index and germination percentage, conducted in accordance with the standard procedures recommended by the Rules for Seed Analysis (BRASIL, 2009).

The water content of samples is used to check the weight loss of seeds when they are subjected to drying processes. Its results guide the adaptation of procedures that contribute to the preservation and quality of seeds. In this study, the water content of araticum-cagao, graviola and pinha seeds was high according to the treatments, making them recalcitrant, i.e. seeds with a high degree of moisture because they had not gone through the dehydration phase at the point of physiological maturity. This phase is characteristic of orthodox seeds, which tolerate long periods of desiccation. Recalcitrant behavior reduces the possibility of seeds desiccating or germinating in the dry stage. A high percentage of water can trigger deterioration and loss of seed vigor.

Electrical conductivity acts as a facilitator in the selection of useful seeds to be germinated, as it allows cell membrane integrity to be assessed indirectly by quantifying leachates in the seed soaking solution. It was expected that the phytohormone would act as a protector, but the results showed that there was no contribution from GA_3 to the reorganization of cell membranes, compared to the control, except for treatment 4, which showed a negative effect on araticum-cagao and pine cone seeds. For graviola seeds, phytohormone acted unfavorably, with greater loss of organic substances to the soaking solution in treatment 4.

The germination test is useful for determining the value of a batch of seeds, their sowing rate in the field, comparing batches and marketing seeds. However, as germination is a complex process, carrying out the test in the field is made difficult by varying environmental conditions, especially the occurrence of different temperatures and substrates.

Each species has an optimum temperature for the germination of its seeds, and this can be verified by the maximum amount of a batch of seeds that have germinated. Soaking, which is the first stage of germination, requires the right temperature to accelerate the rate of water absorption and the resulting biochemical reactions, resulting in greater uniformity in the process.

The species' preference for the type of substrate also contributes to their development. Choosing the right substrate requires knowledge of the species' relationship with water and light, as there are seeds that are positive photoblasts, germinating only in the presence of light, others that are negative, germinating in the absence of light, and others that are insensitive to light, known as neutral.

In laboratory conditions, external factors can be controlled, favoring complete, rapid germination and producing uniform seeds.

The germination test is considered in all its phases, starting with soaking, through the activation of metabolic processes and the development of the embryo until it reaches the plantula stage, still dependent on the seed's reserves.

The germination percentage test corresponds to the proportion of the number of seeds that produced normal seedlings.

In this study, for an unknown reason, no germination was seen in araticum seeds obtained commercially; germination did occur in those collected in the field and the contribution of GA_3 , at a concentration of 1000 mg L^{-1} was seen, in the presence of the substrates germitest paper and sand plus topsoil. For graviola seeds, there was a contribution from GA_3 , at a concentration of 500 mg L^{-1} in combination with germ paper and sand plus topsoil. It should be noted that GA_3 at a concentration of 2000 mg^{-1} drastically affected the process, with no seeds germinating. For pine cone seeds, GA_3 contributed at concentrations of 200, 400 and 800 mg L^{-1} when combined with the substrates sand plus topsoil, cerrado topsoil and sand.

It was therefore observed that the substrates: sand plus vegetable soil and cerrado soil stood out for having a direct influence on germination, providing water and oxygen thanks to their structure and aeration, as well as offering physical support for the development of the seedlings.

However, this study is the starting point for understanding and/or facilitating the germination of araticum-cagao, graviola and pine cone seeds.

REFERENCES

BRAZIL. Ministry of Agriculture, Livestock and Supply. Secretariat for Agricultural Defense. **Rules for Seed Analysis**. Brasilia: Mapa/ACS, 2009. 399 p. ISBN 978-85-99851-70-8

CARVALHO, P. E. R. **Species of Brazilian trees**. Brasilia: Embrapa Informagao Tecnologica, v. 2, 2003.

LEMOS, E. E. P. de. Anonaceae production in Brazil. Jaboticabal-SP, **Revista Brasileira de Fruticultura**, v. 36, n. spel, p. 77-85, 2014.

MIX
Papier aus verantwortungsvollen Quellen
Paper from responsible sources
FSC® C105338

Printed by Books on Demand GmbH, Norderstedt / Germany